50 Science Numbers Everyone Should Know

Richard J. Fruncillo, MD, PhD

Disclaimer

I have made many attempts to ensure the accuracy of the information contained in this book at the time of publishing. However, because science is a rapidly changing discipline, neither the author nor the publisher can guarantee that the information in this book is complete and accurate. There is also the possibility of human error. Neither the author nor the publisher can be responsible for damages that are perceived to result from the use of this book.

Introduction

I have always been fascinated by science. I became a lifelong student after an encounter with an eight-chemical chemistry set around 1960. This second decade of the twenty-first century is an extremely exciting time for scientific knowledge. We are close to having an understanding of all the major questions humans have pondered for thousands of years. From the origin of the universe to the origin of life, from the fundamental forces of nature to the creation of the elements, and from the first humans to an understanding of human diseases, we are finally at a point where science can apply plausible, unifying principles to all of these areas.

However, despite the above, it sometimes seems that interest and enthusiasm for science may be at an all-time low. In this fast-paced, information-and-image-saturated era, society seems to be focused on the latest technology with little interest in how we arrived at this point. I am always shocked when I show people a real-time view of Jupiter or Saturn through a twenty-inch telescope, and the response is, "Is that it?" I am equally disillusioned when I have a failed attempt to initiate a discussion about one of the topics in this book with a colleague who holds a college degree in science.

The purpose of this short book is to make the average nonscientifically inclined person, or a former scientifically inclined person, capture some of this twenty-first-century enthusiasm. I have oriented this book around numbers because I think they provide the most fascinating and compelling examples of a particular topic. As you will see in the text, some of the numbers are still in question.

Newer developments in the social and behavioral sciences are not presented because of space considerations. As a broad summary, scientists in these areas postulate that many of our current behaviors, including mate selection, group loyalty, overreaction, etc., stem from evolutionary mechanisms to protect and reproduce our species. One area that could be considered the last remaining frontier of scientific research is that of the origin of consciousness. To date there is no good scientific explanation for it.

The reader should begin by looking at all fifty questions posed at the beginning of the book. He or she should then think about what the answer to the question may be. The reader should then read each question, answer, and explanation. I guarantee that many readers will be astonished by some of the answers and how far we have come in our understanding of these questions. The reality is that many of the questions asked in this book could not have even been asked fifty to one hundred years ago because the intellectual basis for the question did not exist.

Enjoy yourself, and know that if you develop an understanding of the numbers and explanations in this book, you will have a knowledge base for most of the major scientific issues of the present. You might also have some wonderful new material for small talk at cocktail parties.

Powers-of-Ten Notation

Some of the numbers in this book are either too big or too small to be conveniently written. Such numbers will be expressed in scientific—or powers-of-ten—notation. Any number can be written as the product of a coefficient that is between 1 and 10, and 10 to an integer power. For numbers less than 1, the power of ten becomes negative. The coefficient is usually expressed to a few significant figures. Examples are below:

Number	Scientific notation
10	1×10^1
1	1×10^0
424,000	4.24×10^5
756,000,000,000	7.56×10^{11}
0.000258	2.58×10^{-4}
0.00000001678	1.678×10^{-8}

A Word about Units

Many of the numbers in this book are expressed in the familiar terms of years and miles. Temperature is expressed both in the Celsius and Fahrenheit scale. Water freezes at 0°C and 32°F. It boils at 100°C and 212°F. Very small distances are stated in the metric system in fractional meters. The following prefixes are used:

Milli	one-thousandth	(10^{-3})
Micro	one-millionth	(10^{-6})
Nano	one-billionth	(10^{-9})
Pico	one-trillionth	(10^{-12})
Femto	one-quadrillionth	(10^{-15})

Large numbers are expressed with the following prefixes:

Kilo	one thousand	(10^{3})
Mega	one million	(10^{6})

The unit of energy used is the joule (J), which is the energy expended when a force of one newton is applied over a distance of one meter. Stated another way, it is approximately the energy required to lift a small apple a vertical distance of one meter. The metric unit of volume is the liter, which is 1000 cubic centimeters or 1.06 quarts.

Questions

1. How old is the universe?

2. How many universes are there?

3. How old is the earth?

4. What is the distance from the sun to the edge of the solar system?

5. How many exoplanets (non-solar-system planets) have been discovered?

6. What is the speed of light?

7. What is the distance to the nearest star outside our solar system?

8. How many stars are in our galaxy?

9. How many galaxies are in the universe?

10. What percentage of the universe is dark matter?

11. What percentage of the universe is dark energy?

12. How dense is the center of a black hole?

13. How much energy is released in a gamma ray burst?

14. When did life first appear on earth?

15. When did multicellular life first exist?

16. When did our ancestors first walk erect?

17. When did modern humans first appear on earth?

18. What is the size of a eukaryotic cell?

19. What is the typical size of a bacterium?

20. What is the average size of a virus?

21. What is the weight of an ant's brain?

22. How many genes do humans have?

23. What is the number of living species?

24. When did the dinosaurs become extinct?

25. What is the average human life span?

26. How many times a day is a human cell at risk for development of cancer?

27. When do organs begin to form in a human fetus?

28. What is the safe minimum temperature to cook hamburger meat?

29. What is the wavelength of yellow light?

30. What is the smallest image that can be seen with an optical microscope?

31. What is the average radius of an atom?

32. How many places can an electron be around an atomic nucleus?

33. What is the size of an average atomic nucleus?

34. What is the size of a quark?

35. What is the size of a string?

36. What was the amount of excess matter compared to antimatter formed at the big bang?

37. How many elementary forces are there?

38. What is the temperature of the hottest fire?

39. What is the temperature at the center of the sun?

40. How many chemical elements are there?

41. How much uranium is required to make a nuclear weapon?

42. What is the lowest temperature?

43. At what temperatures do rocks melt?

44. How much does a continent move in a year?

45. How deep is the ocean?

46. What percentage of the world's freshwater is stored in glaciers?

47. How deep is Antarctic ice?

48. How many years did it take to produce the world's oil and coal supply?

49. What is the energy density of gasoline?

50. How many transistors can be placed on a computer chip?

Q 1. *How old is the universe?*
A. 13.7 billion years

The most commonly accepted scientific explanation for the origin of the universe is the big bang theory. It postulates that the universe started at time zero as an incredibly hot, extremely small bundle of pure energy. Over time this packet of energy expanded and cooled, and according to Einstein's famous equation $E = mc^2$ (which teaches that mass and energy are interchangeable) it eventually transformed into particles with mass that make up atoms. A force known as inflation jump-started the universe into a process of massive expansion. Due to nonuniformities in regional distributions of matter, the atoms became subject to the force of ordinary gravity, and coalesced into stars, planets, and galaxies.

With today's powerful telescopes, scientists can measure the intergalactic distances and speed of expansion between galaxies. If both of these values are known, a calculation can be performed to determine the time at which both objects began as a single point. This process yields the result 13.7 billion years, which is considered the age of the known universe. To use an example, if a bullet travels at a velocity (V) of 1000 feet per second, and it hits a target at a distance (D) of 500 feet away, we can calculate the time that the bullet left the gun as D/V or 0.5 seconds. This method of calculation is really an oversimplification, as it only applies if the expansion of the universe is constant over time. Scientists believe that the expansion of the universe is not constant, but that it is actually accelerating by a process known as *dark energy*. In the distant future, scientists believe, other galaxies will move away from us at the speed of light, and we will never see them.

Q 2. *How many universes are there?*
A. Possibly one, possibly many

At first this question seems a logical contradiction. The definition of a universe is *that which contains everything*. So how could there be more than one? The reason to entertain such a question stems from a paradox of nature that involves the physical constants that govern the behavior of the universe and that allow for the development of the complicated structures and processes within it. For example, if the gravitational constant (the number that determines the force of gravity) or Planck's constant (the number that determines the behavior of subatomic particles) were just a little different, the universe as we know it—consisting of stars, galaxies, elements, and life—would not exist. To some, it seems highly improbable that values for all of these constants of nature developed by chance. They postulate that our universe is one of many universes, each with its own unique set of physical constants and properties. Some of these other universes may be all hydrogen or all black holes, or they may harbor a completely different form of intelligent life.

Some cosmologists believe that there is a theoretical basis for all of the universes of the multiverse arising from nothing. With mathematical reasoning they postulate that all universes started with the creation of a huge amount of positive mass-energy and a huge amount of negative gravity. This negative gravity fueled inflation. The sum of the two equals zero, or the original nothing.

Although there is some theoretical basis in string theory and inflation theory for the existence of a multiverse, many scientists do not even consider this a scientific theory since it seems impossible to prove or disprove by experiments. So for now the best answer to the above question is one!

Q 3. *How old is the earth?*
A. 4.54 billion years

The earth was formed at the same time as the sun and other planets of our solar system. It is believed that at a time when our universe was about nine billion years old, a large dust cloud began to condense on itself in a process that would lead to the formation of our sun. As the particles became closer together by the force of gravity, they caused the entire cloud to spin and flatten into a disc. The central core of this spinning system eventually became our sun. A small fraction of the remaining material condensed to form the planets. Support for this theory comes from the fact that all the planets revolve around the sun in roughly the same plane, and most have the same spin direction.

The age of the earth is calculated by the method of radiometric dating applied to rocks on the earth and to meteors. An example of this is potassium-argon dating, which measures the formation of argon—an inert gas—from an isotope of potassium. When a rock is first formed, it solidifies from a liquid state. While a hot liquid, any argon gas present will boil away. As a solid rock, argon produced by radioactive decay will be trapped inside the inner matrix of the rock. By knowing the decay rate and the ratio of potassium to argon present, scientists can calculate the age of the rock. Applying this method to rocks on earth can be problematic since the earth is stratified into core, mantle, and crust, and older rocks can be contaminated by residue from newer rocks. However, certain meteors are considered to represent remnants from the original formation of the solar system. Applying the dating process to them, in combination with data obtained from terrestrial rocks, yields a value of 4.54 billion years for the age of the earth.

Q 4. *What is the distance from the sun to the edge of the solar system?*
A. 50,000 AU or 4.65 trillion miles

Dimensions within the solar system are usually specified in terms of astronomical units (AU). One AU is the distance from the sun to the earth or 93,000,000 miles. Since Pluto was demoted as a planet in 2005, one would think that the solar system ends with the orbit of Neptune—roughly 35 AU. However, we now know that Pluto is one of possibly over one thousand icy objects that orbit in a region between 30 and 50 AU known as the Kuiper belt. These objects can interact gravitationally with the outer planets and can be the source of comets that orbit the sun less than every two hundred years.

When comets come close to the earth, they appear as a bright disc with a long tail. Kuiper belt objects have orbits that can be circular or oval and are close to the plane of conventional planet orbits. Recently astronomers have hypothesized the existence of a spherical collection of objects called the Oort cloud located up to 50,000 AU (one-fifth of the distance to the nearest star). These bodies are too faint to be seen with current telescopes and are composed mainly of frozen ammonia, methane, and water. They are thought to have been hurled out in a direction perpendicular to the plane of the solar system by the gravitational forces of the large planets. At 50,000 AU the sun would appear as just another star to a person standing on one of these objects. The Oort cloud is postulated to be the source for the long-period comets (comets that orbit the sun greater than every two hundred years). The outer boundaries of the Oort cloud represent the absolute limit of the sun's gravitational force and define the beginning of interstellar space.

Q 5. *How many exoplanets (non-solar-system planets) have been discovered?*
A. Over nine hundred, but trillions may exist

The traditional definition of a planet is a spherical object that orbits the sun, shines by reflected light, and has a well-defined orbit. For many years scientists have considered the existence of planetary systems around other stars, but the technology did not exist to detect them. In 1995 the presence of the first non-solar-system planet, or exoplanet, was confirmed by measuring the gravitational wobble in its companion star. Since then, over nine hundred exoplanets have been found, and new ones are being discovered at a rapid rate. Some methods of detecting exoplanets depend on the effect of gravity, while others measure the variability of the light intensity of the parent star as the planet crosses its path. Low-resolution photos of exoplanets do actually exist.

Most extrasolar planets that have been discovered are larger than Jupiter and are close to being ministars. However, since it is easier to detect these large planets, there is a sampling bias toward them. Many have noncircular orbits, and many have orbits close to their parent star, which definitely places them outside the habitable zone for life as we know it. However, at the time of this writing, there is news of the discovery of two Earth-sized exoplanets.

Many stars with planets contain multiple planetary systems. Some astronomers are now willing to conclude that planetary formation is a normal part of star formation and that there are probably one to ten billion, or more, planets orbiting stars the size of our sun in our Milky Way galaxy. The number of planets in the universe could be in the trillions! How may have life as we know it?

Q 6. *What is the speed of light*?
A. 186,000 miles per second

Humans have pondered the nature of light throughout the ages. Is it instantaneous, or does it have a fixed speed? During the latter part of the nineteenth century, Maxwell predicted that light should have a finite velocity of 186,000 miles per second in a vacuum. At the turn of the twentieth century, Albert Einstein postulated in his special theory of relativity that the velocity of light is constant, regardless of your frame of reference. Our experience tells us that if a person at rest on earth observes another person throwing a ball at thirty miles per hour from a car moving at fifty miles per hour, the ball will appear to him or her to move at eighty miles per hour. However, according to relativity, light from a flashlight turned on from a car moving at one-half the speed of light will still appear to move at 186,000 miles per second from a fixed reference point.

The theory of relatively has many strange consequences. When an object approaches the speed of light, the perception of time on the object will slow down, the length of the object will shorten, and the mass of the object will increase. A person traveling at the speed of light will perceive time as standing still. Nothing in the universe can move faster than light because it would be moving back in time. A mathematical deduction of relativity theory is the famous equivalence relationship between mass and energy, $E = mc^2$, where E is energy, m is mass, and c is the speed of light. These bizarre predictions have been confirmed by experimental evidence many times, and relativistic effects have to be taken into consideration in the engineering of modern electronic devices

Q 7. *What is the distance to the nearest star outside our solar system?*
A. 4.3 light-years or 25.4 trillion miles

A star is defined as a large gaseous body that has a core temperature hot enough to allow for nuclear fusion of hydrogen into helium. This is in contrast to gas-giant planets like Jupiter and Saturn, which are not large enough to have sufficient gravitational energy in their centers to produce the heat necessary to begin thermonuclear reactions. Stars come in varying sizes, and the larger ones are capable of nuclear fusion of other elements besides hydrogen.

Distances outside the solar system are measured in terms of light-years, or the distance it takes light to travel in one year (approximately 5.9 trillion miles). The star closest to the earth is obviously our sun, but the star closest to the earth outside our solar system is called Proxima Centauri. It is part of a three-star system consisting of itself and a binary star (two stars rotating around each other). It is believed that Proxima Centauri is slightly closer to the earth by about 0.1 light-years than the other two binary star members of the system.

Although this is the closest star, it impossible to imagine with current technology how any spacecraft would ever be able to travel to it. The United States is currently sending a probe to Pluto, which was launched in 2006. It will take the spacecraft 9.5 years to travel the 3.07 billion miles to Pluto. At that rate, travel to Proxima Centuri would take approximately 78,600 years one way!

Q 8. *How many stars are in our galaxy?*
A. Two hundred billion

A galaxy is a huge collection of stars, gas, and dust organized into a characteristic shape and held together by gravitational forces. Galaxies can be spiral, elliptical, or irregular. Recently, scientists have calculated that the visible matter in galaxies is not enough to allow them to maintain their stability, and have postulated the existence of an exotic form of matter called *dark matter*. It is thought that this material surrounds galaxies in a type of halo and maintains their structural integrity.

We live in an average-sized galaxy called the Milky Way, consisting of about two hundred billion stars. It is estimated that larger galaxies can contain as many as ten trillion stars, while smaller ones might have as few as ten million stars. The center of the Milky Way contains a supermassive black hole—an object that is so dense that it is invisible because light cannot escape from it. It is thought that most galaxies contain large black holes in the center to form the basis for their structural organization.

An important event in galaxies is continued star formation from condensation of interstellar gas. This process was much more active in the early universe. At present, most galaxies create about one new star per year. However, simulations have shown that when galaxies collide, the rate of star formation can increase by a factor of one hundred. As stars age, depending on their initial size, they collapse into white dwarfs, neutron stars, or black holes. Some stars explode in a spectacular event called a supernova. Matter from these stellar explosions is dispersed into space and provides material for future star formation.

Q 9. *How many galaxies are in the universe?*
A. One hundred billion

Like stars, galaxies can group together and form binary (two objects rotating around each other) or higher-order systems. The Milky Way galaxy is part of a small group of galaxies known as the Local Group, which spans a distance of a few million light-years. Small groups of galaxies can be a part of a larger system known as a galaxy cluster. A well-known cluster is the Virgo cluster, which contains over one thousand individual galaxies. Amazingly, clusters of galaxies can band together to form superclusters, which can be as large as one hundred million light-years in length.

Galaxies do not remain static but interact with others nearby. In time they will change shape and merge with each other. The Hubble Space Telescope has provided dramatic images of galaxies colliding and engulfing each other. In the Hubble Deep Field Photo, galaxies from 13 billion years ago are visible. These pictures demonstrate that galaxy collisions were much more common in the early universe. In the future a large cluster of galaxies might eventually end up as two or three supergalaxies.

At present, it is thought that the universe is composed of a three-dimensional weblike network of galaxies separated from each other by voids (large regions of empty space). Astronomers estimate that there are approximately one hundred billion galaxies in the known universe.

The physics of galaxy formation and evolution is not well understood, but computer simulations have predicted the formation of superclusters and voids. These simulations are also consistent with visual observations and support the existence of dark matter.

Q 10. *What percentage of the universe is dark matter?*
A. 24% of the mass-energy

If we total all the energy in the universe, then add to this number the energy equivalent of all the mass in the universe (by $E = mc^2$), we discover that dark matter makes up about 24% of the total. It seems shocking that dark matter is all around us, and we don't know much about the particle associated with it. We do know that it originated in the big bang, it is probably massive, and it only interacts gravitationally with atomic matter (neutrons, protons, and electrons).

The existence of dark matter is inferred from observations of galaxies. Radio telescopes have demonstrated the presence of large rotating discs of hydrogen gas beyond the visible part of the galaxy. The rotation speed of these discs is consistent with the presence of more galactic mass than can be accounted for by the mass of all the visible stars. Similarly, galaxy clusters rotate at speeds that imply that they are much more massive than the sum of their individual star masses. Also, the immense mass of galaxies can actually bend light from other distant galaxies behind them. This effect is called gravitational lensing. By measuring the degree of light distortion, scientists have concluded that galaxies have much more mass than what is visually apparent.

It is thought that the purpose of dark matter is to give the universe definition. Without it, atomic matter may not have organized into stars and galaxies. In the early universe, atomic matter was evenly distributed, but dark matter had some "lumpiness" due to quantum effects of inflation. In time gravity condensed the dark matter, which then attracted the atomic matter—thus the origin of stars.

Q 11. *What percentage of the universe is dark energy?*
A. 72% of the mass-energy

If you were astounded at the amount of dark matter in the universe, you will be more astounded by the fact that the sum of dark matter (24%) and dark energy (72%) accounts for 96% of the mass-energy of the universe; the familiar atomic matter and energy (neutrons, protons, electrons, photons) accounts for 4%.

Dark energy is even more mysterious than dark matter. It is an energy that is present uniformly everywhere, even in the vacuum of space. It is not diluted by the expansion of space, and in a weird way it causes an acceleration of space. It is thought that it did not have a role in the early universe, but that about nine billion years after the big bang, it began to play a major part in the accelerating expansion of the universe. It acts as an opposition to gravity, and as such opposes the formation of structure. Thus the future universe will contain large areas of emptiness.

Two theories for dark energy are: (1) it is a constant energy associated with space itself, or (2) it can be another form of energy associated with a new field in nature that remains almost constant as the universe expands but varies very slightly. The latter theory can be tested since a variable dark energy field means a different rate of expansion compared to that of a constant dark energy field. In the future it may be possible to experimentally confirm that there is or is not a difference and thus lend support to one idea over the other.

Q 12. *How dense is the center of a black hole?*
A. Infinite

Near the end of a star's life span, thermonuclear reactions cannot oppose the effects of gravity. Stars the size of the sun will contract to the size of the earth and continue to emit energy from non-fusion-related changes in energy levels of atomic nuclei (white dwarf). Eventually, the atomic nuclei will cool to the point where they no longer emit visible light, and the white dwarf will change into a black dwarf. A teaspoon of white dwarf matter would weight over a ton.

A neutron star is formed when the core of a very large star implodes, producing forces that are so great that electrons are pushed into the protons of atomic nuclei. The resultant object is a mass of all neutrons that rotates very quickly, has a strong magnetic field, and produces periodic radio waves in the form of pulses. A teaspoon of material from a neutron star would weigh approximately a billion tons.

Neutron stars greater than three to five times the mass of the sun can further collapse into black holes. The mass density of these objects is so extreme that no light can escape, space and time are distorted, and time near a boundary called the event horizon would appear to slow down to an outside observer. The mass of a black hole is mathematically considered to be concentrated at a point in the center known as a singularity. Theoretically, there is infinite density at this point. The actual density will depend on the effects of quantum gravity, which are not known and are a work in progress. However, even though the mass is concentrated in the center at an infinitesimal point, black holes are considered to have finite boundaries defined by the event horizon. Some think that rotating black holes can be *wormholes*, or gateways to other universes.

Q 13. *How much energy is released in a gamma ray burst?*
A. 10^{44} joules (energy of 10^{29} H-bombs)

When a rapidly rotating large star explodes into a supernova, an extremely high-energy emission takes place as the star collapses into a neutron star or black hole. Known as a gamma ray burst, the occurrence is the brightest and most energetic event in the universe. Bursts can last from milliseconds to minutes but typically last seconds. The initial burst of gamma rays is usually followed by periods of electromagnetic energy emission at longer wavelengths. They are extremely rare and postulated to occur at the rate of 3–4 per galaxy per million years. This translates to one event per day somewhere in a far off galaxy.

The energy from these bursts is ejected in a narrow beam similar to a laser. They are thought to result from collisions of particles moving at relativistic velocities producing photons with very short wavelengths. As the photons interact with gases in the star, they produce the longer wavelength afterglow emissions in the X-ray, visible, and radio-wave areas. If the energy of a gamma ray emission was dispersed spherically, one emission could have the energy equivalent of the mass of two suns released in seconds. Since the energy is discharged in beams, the calculated value for the release is approximately 10^{44} joules, or the mass energy of 1/2000 of the sun's mass, or the energy of 10^{29} hydrogen bombs.

Gamma ray bursts from distant galaxies are harmless to life on earth. However, one originating in our own galaxy pointing at us could be devastating. It has been postulated that several mass extinctions were the result of nearby high-intensity bursts.

Q 14. *When did life first appear on earth?*
A. 3.85 billion years ago

Life is defined as a self-replicating chemical system that extracts energy from the environment and is capable of evolution. There are many theories for the origin of life. Most believe that life originated in the oceans, perhaps near pockets of volcanic activity on the ocean floor known as hydrothermal vents. The first life forms were probably structurally similar to modern-day bacteria or archaea. Though similar to bacteria, scientists believe that there are enough differences between bacteria and archaea to declare archaea a separate domain. The early life forms developed in an earth environment that had a very low oxygen concentration in the atmosphere. Because of this, they needed an energy source that wasn't oxygen dependent. Their biochemistry likely resembled that of *extremophile* archaea that exist at high temperatures and obtain their energy supply from inorganic material.

The first fossil evidence for life dates back to approximately 3.5 billion years, but radioisotopic methods have revealed evidence of life as early as 3.85 billion years ago. One form of early life was similar to modern-day blue-green algae, and was thought to be responsible for increasing the atmospheric concentration of oxygen on the earth. A sudden increase of iron oxide on rocks supports this.

The two necessities for life are (1) a group of chemical reactions capable of taking energy and precursor molecules from the environment and converting them to structure and (2) a self-replicating method of transmitting genetic information. It is debated whether one process came first or they coevolved. It is also debated whether archaea evolved from bacteria or vice-versa, or whether both evolved from a more primitive common ancestor.

Q 15. *When did multicellular life first exist?*
A. Six hundred million years ago

Around 2.5 billion years ago, oxygen concentrations in the atmosphere had reached such high levels that it became toxic to many existing prokaryotes. These organisms then evolved into those that relied on oxygen for energy. Around two billion years ago, prokaryotic cells merged to form larger *eukaryotic* single-celled organisms. Also at this time, sexual reproduction first appeared, allowing for a mix of genetic information from the two parent cells and ultimately allowing for more diversity.

Sometime around six hundred million years ago, life suddenly erupted into all types of multicellular organisms. This was a very complicated process as single cells now had to organize and communicate with billions of other cells. There is an abundance of fossils representing different forms of multicellular invertebrate life present in rocks dating from 530 to about 570 million years ago. This burst of different life forms is called the Cambrian explosion. Subsequent key events (before present) in the evolution of life were:

1. First wormlike vertebrates—500 million years
2. Plants and arthropods colonize the land—475 million years
3. First vascular plant—420 million years
4. First vertebrates leave the oceans—400 million years
5. First animal with four legs (tetrapod)—395 million years
6. First reptiles and first forests—350 million years
7. First primitive mammals and first dinosaur—220 million years
8. First flowering plants—125 million years
9. First marsupials—110 million years
10. First snakes—95 million years
11. First primates—55 million years
12. First primitive whales—50 million years
13. First monkeys—45 million years
14. Separation of apes and humans—5 million years

Q 16. *When did our ancestors first walk erect?*
A. 4.4 million years ago

The question is not that simple. Apes can walk on two legs (bipedality) for short periods of time, but their skeletons are not adapted to constant bipedality like humans' skeletons are. Primates are called hominins if they are more closely related to humans than to apes. There was a primate named *Ardipithecus ramidus* that existed about 4.4 million years ago. It had hand bones that were uncharacteristic of a knuckle walker and had a spinal column that was more vertical than that of most apes. Most scientists consider *Ardipithecus* a hominin better capable of bipedality compared to other primates of the period.

The *Australopithecus* genus composed a varied group of hominins that existed 1.5 to 4 million years ago. They were bipedal, although they still may have taken shelter in trees. They may have used simple tools, but there is no evidence of creative technological ability. The best-known example is the fossil known as Lucy. She was about forty inches tall, weighed about eighty pounds, and lived about 3.2 million years ago. Her bones demonstrated traits similar to modern humans with respect to bipedality. Her spinal cord entered her brain below the skull (in quadrupedal primates, the spinal cord enters the brain through the side of the skull). However, her brain was not large. It was 450 cc, which is about 25% larger than a chimpanzee's brain.

There are 3.6-million-year-old hominin fossil footprints from a presumed *Australopithecus* species that appear to represent two distinct individuals walking side by side. There is spacing between the footprints to suggest that a larger hominin was walking slowly to keep pace with a smaller one. Some scientists believe that these footprints have characteristics that are almost identical to the feet of modern humans.

Q 17. *When did modern humans first appear on earth?*
A. 190,000 years ago

About 1.5 to 2 million years ago, *Homo habilis*, *Homo erectus*, and an advanced species of *Australopithecus* cohabited the earth. The *Homo* genus was clearly becoming superior as brain size, body size, and a larynx capable of speech provided a significant survival advantage over the australopithecines, which disappeared about 1.5 million years ago.

Homo habilis is considered the first hominin to use precisely manufactured tools. They were made from quartz, flint, or obsidian, and were rubbed together to create sharp edges capable of cutting scavenged meat. *Homo habilis* had a brain size of 650 cc, a more vertical face, and smaller teeth compared to *Australopithecus*.

The earliest *Homo erectus* appeared about 1.9 million years ago. Its average brain size was 900 cc, it ate more meat, its tool kit was advanced, it was the first *Homo* to leave Africa and walk all over the world, and it may have controlled fire. The clear advantages of *erectus* over *habilis* eventually led to the extinction of *habilis* and the evolution of *erectus* to the *sapiens* lineage.

The first *Homo sapiens* appeared about 190,000 years ago, and are considered to be early modern humans. They had a brain size of 1250 cc, a flattened face, and absent brow ridges. About 150,000 years ago, the Neanderthals appeared in Europe. Modern genetic studies have shown that Neanderthals had some genes similar to modern humans and that they possibly interbred with modern humans. They disappeared about 30,000 years ago for unknown reasons—probably because they were outcompeted by *Homo sapiens* in areas such as technology, language, hunting, and disease susceptibility.

Q 18. *What is the size of a eukaryotic cell?*
A. 10 micrometers or 10^{-5} meters

The eukaryotic cell is the basic building block of all multicellular plants, fungi, and animals. It is thought to have first appeared about 2 million years ago from a merging of smaller and less complex prokaryotic cells. There is some biomarker evidence that eukaryotes existed as early as 2.7 million years ago.

The essential element that distinguishes a eukaryotic cell from a prokaryotic (bacterial) cell is the presence of a cell nucleus that contains the genetic material. Reproduction in eukaryotes takes place by mitosis (each daughter cell receives the same number of chromosomes) or meiosis (daughter cell receives half the number of chromosomes).

Other key organelles of eukaryotes are mitochondria, chloroplasts, and Golgi apparatus. Mitochondria convert high-energy organic molecules into the energy-unit molecule ATP, which provides the energy for cellular work. Chloroplasts are found in plants and photosynthetic microbes. They convert the energy of sunlight into sugars. The Golgi apparatus is important in storing and processing macromolecules for secretion. Both mitochondria and chloroplasts have their own DNA and can reproduce independently of the parent cell. These facts support the theory that mitochondria and chloroplasts were once independent organisms.

Plant cells differ from other eukaryote cells by the presence of a rigid cell wall. Other structures differentiating eukaryotes from prokaryotes are the presence of microtubules, lysosomes, endoplasmic reticulum, nucleolus, and multiple chromosomes.

The size of a eukaryotic cell is typically stated as being from two to one hundred micrometers, although some neuron cells can be several meters in length.

Q 19. *What is the typical size of a bacterium?*
A. 1 micrometer or 10^{-6} meters

Bacteria are the most abundant living things on earth. They are present in soil, oceans, decaying organic matter, and the intestinal tracts of animals and insects. Their size varies from 0.5 to 2 micrometers, and they can be spherical, spiral, or rodlike in shape. While some are associated with serious infectious diseases in humans and animals, most are harmless or beneficial. Some require oxygen to survive (aerobic), while others do not (anaerobic).

The genetic material of bacteria differs from the more complex eukaryotes by the presence of a single long strand of circular DNA within the cytoplasm of the cell. They have cell walls that can be distinguished by their ability to take up *Gram stain*. They are capable of rapid reproduction with some doubling in number every twenty minutes. They can also exchange genetic information between cells in the form of plasmids. Because of these properties, they are subject to a high rate of mutation and eventual antibiotic resistance. They reproduce by cell division, budding, or fragmentation. Some are capable of forming long-lasting spores. Some scientists have claimed to revive bacterial spores that were millions of years old.

The archaea are a fascinating type of prokaryote that evolved specific structural and biochemical properties that allowed them to flourish in extreme environments. They can be present in 90°C hot springs and highly acidic (pH 1) or high-salt-content settings. Since they thrive in conditions toxic to most other forms of life, they can be hard to culture and observe. Their existence supports speculation about the presence of nontraditional life forms in other areas of our solar system. Some of the more common types of archaea produce methane from carbon dioxide and hydrogen, or metabolize inorganic sulfur. Though originally thought to be a subtype of bacteria, modern genetic analysis surprisingly places the archaea closer to eukaryotes than to bacteria.

Q 20. *What is the average size of a virus?*
A. 100 nanometers or 10^{-7} meters

Viral particles are considered nonliving entities (there is some debate about whether they should be considered alive) that can only reproduce inside living cells. Some can be crystallized like chemicals. They are made up of a core of either DNA or RNA surrounded by a protein coat. Plants, animals, bacteria, and archaea can all be victims of a viral infection. Their size can vary from 20 to 300 nanometers. There has never been fossil evidence of viruses, so their origin in relationship to prokaryotic and eukaryotic life is somewhat speculative.

There have been three popular theories for the origin of viruses:

> 1. Regressive hypothesis. Viruses may have evolved from small cells that were parasites to larger cells. After many years, they lost most cellular components that were unrelated to parasitism. Modern-day chlamydia and rickettsia could be similar to these viral ancestors.
> 2. Progressive hypothesis. Cells came first, and viruses evolved as pieces of genetic material and protein that escaped from the parent cell. The current behavior of plasmids supports this theory.
> 3. Coevolution hypothesis. Viruses evolved at the same time that cells evolved and were always dependent on them.

Other theories suggest that viruses may have predated cells and may have been self-replicating, or that different viruses arose at different times by various mechanisms. Most scientists at present do not believe that all viruses share a common ancestor.

Prions are 10-nanometer infectious protein particles that are distinct from viruses. They invade cells and cause serious diseases such as mad cow disease. They do this by initiating a chain reaction in the folding of normal cellular proteins into a pathologic shape.

Q 21. *What is the weight of an ant's brain?*
A. 10 milligrams

A common misconception is that the bigger the brain, the more intelligent the organism. The largest brains are found in sperm whales and elephants, and although these animals are considered intelligent, they are not more so than parrots and chimpanzees, whose brain sizes are significantly smaller. There is a better, but not perfect, correlation between intelligence and brain size/body weight ratio. In this regard the shrew has the highest ratio but is obviously not the most intelligent.

It has become increasingly recognized that larger animals need bigger brains just to coordinate the movements of all the muscles necessary for locomotion. As a brain grows larger, it incurs increased infrastructure just to maintain the ability to manage neurons over long distances. Also, energy requirements increase as brain size increases. This latter point may force an evolutionary limit on the future development of the human brain.

The average ant has a 10-milligram brain. It is capable of complex sensory, motor, and interindividual communication. These processes are possible with such a minute amount of neuronal tissue because all of the neural connections are within millimeters of the central brain. Compared to larger animals, the infrastructure and energy requirements of ants and other insects are very low.

The key feature of the ant's brain is the emphasis on odor and pheromone (chemicals released that help animals communicate with each other) processing by sensory stimuli received through its antennae. Evolution has reduced the size of worker ant brains by eliminating any unnecessary parts (e.g., sexual and complex visual centers). It is believed that there is a process of emergence in an ant colony in which the sum of the behaviors dictated by many small brains produces the behavior of a larger, more complex system.

Q 22. *How many genes do humans have?*
A. Twenty-three thousand

All human genetic information is stored on twenty-three pairs of chromosomes within the cell nucleus. During the mid-twentieth century, the chemical structure of DNA was revealed, and soon after, the "one gene, one protein" theory was proposed. A gene is defined as a segment of DNA that has a unique sequence of DNA bases, which codes for a unique sequence of amino acids in a protein. This genetic code is based on a relationship between three out of a possible four DNA bases, and one out of a possible twenty amino acids.

During protein synthesis, the DNA information in a gene is first copied by a single strand of messenger RNA. This messenger RNA then moves to a ribosome. Another type of RNA called transfer RNA carries specific amino acids to the ribosome and binds to the messenger RNA. The amino acids link up to form a specific sequence, which creates a distinctive protein.

In 1990 the US government, along with international partners, initiated the Human Genome Project. Its initial goal was to map out all of the important parts of human genes in an attempt to gain insight into the origins of genetic disease. In 1998 a private company began a similar effort with the hope of patenting some human genes (the US government eventually rejected this idea). Both investigations concluded around the same time in the mid-2000s. Some of the project's most important findings were: (1) there are about twenty-three thousand human genes; (2) we have about the same number of genes as lower life forms; (3) the "one gene, one protein" theory may be wrong—genes may code for multiple proteins and may need help from other DNA; (4) 7% of vertebrate genes are unique; and (5) human genes contain areas of repetitive DNA.

Q 23. *What is the number of living species?*
A. 8.75 million, not including bacteria or archaea

A species is defined as a group of living organisms with the ability to mate and produce offspring that are fertile. This definition obviously cannot apply to nonsexually reproducing bacteria and archaea. For these, scientists need to measure similarities in DNA and have chosen an arbitrary limit of 98.7% common DNA as a criterion for prokaryotes to belong to the same species.

There are many practical issues in trying to arrive at an accurate number of total species. It is not easy to tell if one organism would mate with another, and scientists in different parts of the world might describe the same organism as two different species. Presently, about 1.7 million species of plants and animals have been classified (not including prokaryotes); however, many more exist that have not been classified. The largest category is insects, of which about one million have been described. Other approximate numbers are:

Vertebrate animals	62,000
Invertebrate animals	1,300,000
Total plants	321,000
Mushrooms, lichens, brown algae	51,500

It was assumed that the total number of eukaryotes was somewhere between 5 and 100 million. Scientists have recently developed a mathematical model that predicts there are 8.75 million eukaryote species on earth. The number of prokaryote species is highly debated. Leading scientists have given estimates of 100,000 to 10 million. However, because of easy genetic exchange, others consider it meaningless to talk about individual bacterial species. One group has estimated the mass of all the prokaryotes on earth to be on the order of 10^{17} grams.

Q 24. *When did the dinosaurs become extinct?*
A. Sixty million years ago

A significant evolutionary development that occurred about 310 million years ago was the creation of the enclosed external egg. Previously, amphibians were dependent on a water environment to reproduce, and thus could never really stray far from bodies of water. This new class of organisms known as reptiles had evolutionary advantages in the new terrestrial forests that were evolving at the time.

About 250 million years ago, the first dinosaurs appeared. They subsequently evolved into many species in different sizes, ranging from two to two hundred feet long. About five hundred genera of dinosaurs have been identified up to the present.

During the late 1990s, it was definitely proven that birds evolved from dinosaurs with the discovery of multiple examples of feathered dinosaur fossils. One type, *Microraptor,* had feathers on its legs and tail. It is thought that feathers developed from scales and were originally ornamental and not related to flight. An animal related to dinosaurs known as pterosaur evolved around 220 million years ago. It is considered the first vertebrate with the ability for powered flight, and it may have been warm-blooded. One giant version known as *Quetzalcoatlus* is thought to have had a wingspan of thirty-six feet.

A mass extinction of dinosaurs and many other animals occurred about sixty million years ago, probably due to the impact of a comet. Dust from the event could have impaired photosynthesis for many years. Temperatures on earth probably fell drastically. In such a situation, only small burrowing animals and insects would have had a survival advantage.

Q 25. *What is the average human life span?*
A. Males 65.0 years, females 69.5 years

Life span is defined as the number of years an organism will live following birth. Life expectancy is defined as the number of remaining years a person will live at a given age. Life spans vary throughout the globe. In countries with high infant mortality, a better indicator of how long people live would be to compare life expectancies at age one or even age five. In some current societies— and in times before modern medicine—average life spans are skewed to the short side by accidents, wars, and childbirth deaths. As an example, a culture with a mean life span of forty-five years may have many people dying before age thirty-five from the above reasons, and then decreased death rates until age sixty. The current average global human life span, according to the United Nations' 2005–10 calculation, is 67.2 years (males 65.0 years, females 69.5 years). Life spans (years) in select countries according to the United Nations are:

Country	Overall	Male	Female
Japan	82.6	79.0	86.1
Switzerland	82.1	79.4	85.1
Italy	80.5	77.5	83.5
United States	78.3	75.6	80.8
Mexico	76.2	73.7	78.6
Russia	70.3	64.3	76.4
India	64.7	63.2	66.4
Afghanistan	43.8	43.9	43.8
Swaziland	39.6	39.8	39.4

Approximate average life spans in colonial America, ancient Rome, and the Bronze Age were twenty-five, twenty-eight, and twenty-six, respectively.

Q 26. *How many times in a day is a human cell at risk for development of cancer?*
A. One million times per day

Through observation of the behaviors of human cancers, along with animal experimentation, it is believed that cancer develops through three distinct stages defined as initiation, promotion, and progression.

Initiation is the process by which DNA is damaged. It can occur by chemicals binding to DNA producing mutations known as DNA adducts, by physical agents such as X-rays and UV light producing strand breaks, or by viruses. It is thought that in any given human cell, initiation of DNA damage takes place at the staggering rate of one million times per day. How can the cell possibly survive this? There are remarkable DNA repair mechanisms in place that patrol up and down the DNA looking for damage and initiating repair. If the repair is defective, there will be a mutated cell with abnormal DNA, and the cell is considered initiated. These cells ultimately can produce a distorted protein. Initiation is not sufficient in itself to cause cancer. Some initiated cells might remain stagnant and never divide again. Others are so damaged that they die, and others may progress into the next stage of neoplastic transformation.

The second stage of carcinogenesis, called promotion, allows for a selective clonal expansion of initiated cells to a pre-neoplastic collection. Agents that are tumor promoters cause an increase in cell proliferation or a decrease in the natural process that triggers the death of an abnormal cell called apoptosis. Promotion can be reversible.

Progression involves the transformation of a pre-neoplastic cell collection into actual cancer. It is thought that another round of DNA damage must take place at this stage. Progression is irreversible.

Q 27. *When do organs begin to form in the human fetus?*
A. Day seventeen, postconception

Most human eggs are fertilized in the fallopian tube. Six to nine days later, the egg implants into the uterus. Approximately 40%–60% of fertilized eggs never implant. Another 20%–30% of implanted eggs eventually dislodge from the uterine wall. The overall first trimester miscarriage rate is approximately 50%–80%.

Once fertilized, the egg is known as a zygote. By the time of implantation, it transforms into a partially hollow ball called a blastocyst. By day seventeen, the first remnants of the nervous system begin to form. The heart forms at day twenty-two, and sexual differentiation occurs at day forty-three. By day fifty-six, all the major organs are formed.

The period of organogenesis from day seventeen to fifty-six is the time during pregnancy when the embryo is most susceptible to the harmful effects of drugs, chemicals, infections, and radiation. Sometimes exposure to damaging agents occurs because pregnancy is not recognized. It is thought that morning sickness is an evolutionary first-trimester adaptation to prevent toxins from being absorbed during the vulnerable period of organ formation.

Drugs and other toxic agents can have detrimental effects on fetal outcome during all three trimesters. Commonly used harmful substances include alcohol, tobacco, and cocaine. It is estimated that approximately 25% of women in the United States do not stop cigarette smoking while pregnant. One study showed that 6%–45% of women presenting to hospitals for delivery had recent cocaine use.

Q 28. *What is the safe minimum temperature to cook hamburger meat?*
A. 160°F

Most *E. coli* bacteria are harmless and colonize the intestines of humans and animals. However, one strain of *E. coli*, O157:H7, present in cow manure has been a source of serious and even fatal human illness. Most cases involve bloody diarrhea and eventually resolve, but children and elderly can develop renal failure and death.

Sometimes at beef harvesting facilities, beef can accidently come in contact with intestinal bacteria containing the pathogenic *E. coli*. In ordinary steak, the outside would be contaminated, and normal cooking procedures that cook the outside of the meat would usually destroy the bacteria. Hamburger meat is particularly vulnerable since the center of a hamburger contains material that once could have been on the meat surface and could have contacted *E. coli*.

In kitchens and restaurants, beef should be handled, stored, and cooked properly. Ground beef, and the utensils used to prepare it, should not contact other foods. Hands should always be washed after touching raw ground beef. Also, all hamburgers should be cooked to a minimum internal temperature of 160°F as measured by an accurate meat thermometer, which should be inserted at several places in the center and thickest part of the patty. Cooking time and external meat color should not be relied upon since the center may be frozen or the outside may be discolored.

E. coli outbreaks have also been associated with the consumption of raw vegetables (usually contaminated with cow manure) and unpasteurized apple cider and juice. Other sources of the disease include contact with farm animals at petting zoos and swimming in contaminated lakes.

Q 29. *What is the wavelength of yellow light?*
A. 570 nanometers

During the nineteenth century, Young and Maxwell demonstrated that light consisted of alternating electric and magnetic fields that propagated through space in waves. The distance between the peaks of waves is called wavelength. The frequency (*f*) of a wave is inversely proportional to wavelength. In the early twentieth century, Einstein theorized that light consisted of particles in his paper on the photoelectric effect. It is now thought that light has both wave and particle properties. As a particle, light is transmitted by a massless entity called a photon.

Visible light is just one part of a whole spectrum of electromagnetic radiation that exists in the universe and is propagated by photons. Quantum theory states that the energy distribution of photons is not continuous. Allowable energies are equal to the product of a very small constant (Planck's constant) and frequency. Examples of electromagnetic waves are:

Type	Wavelength	Frequency (Hz)
Gamma rays	.01 nm	10^{19}
X-ray	1 nm	10^{18}
Ultraviolet	100 nm	10^{16}
Visible light	500 nm	10^{15}
Infrared	10 μm	10^{13}
Microwave	1 cm	10^{10}
Radio/TV	5 m	10^{8}

Visible light has a wavelength range of 400–700 nm and is at the middle of the electromagnetic spectrum. Yellow light is at the middle of the visible spectrum and has a wavelength of 570 nm. White light, along with light from the sun, has light of all visible wavelengths.

Q 30. *What is the smallest image that can be seen with an optical microscope?*
A. 250 nanometers

An optical microscope uses visible light to magnify an image by means of a lens that focuses the light. An early problem with light microscopes was the failure of the lenses to focus light the same way for all the colors in white light. This problem was solved by changing the shape of the glass lens and producing an achromatic lens.

The magnification limit of a microscope depends on the ability to discriminate between two points. This ability is dependent on the wavelength of the electromagnetic radiation used to display the image—the smaller the wavelength, the better the resolution. For the best optical microscope, this lower limit is approximately 250 nm.

To exploit this wavelength relationship, other types of microscopes were invented. The electron microscope uses a beam of electrons as the radiation source for a microscope. Since electrons have short wavelengths, the electron microscope can resolve objects down to 50 picometers (50×10^{-12} meters). They can also be used to see fine detail in larger objects. Instead of a glass lens, magnetic fields are used to focus the electron beams. A disadvantage is that biological specimens require special preparation to be viewed. A microscope using X-rays exists and has the advantage of visualizing biological specimens in natural conditions.

Another type of microscope depends on touch. Known as the scanning probe microscope, the device can record very subtle differences in the surface topography of an object with a probe that can be as sharp as less than a dozen atoms across. Although the resolution and magnification of such an instrument is high, a disadvantage is that it can only visualize an object's surface.

Q 31. *What is the average radius of an atom?*
A. 150 picometers (150×10^{-12} meters)

An atom is the basic building block of chemical elements. It was defined by the Greeks as that which cannot be further subdivided. We now know this statement is not exactly true since atoms are composed of a nucleus of protons and neutrons surrounded by clouds of electrons in distinct orbitals. The nuclear particles can be further subdivided into entities called quarks.

It is hard to define the boundaries of an individual atom since theoretically there is no location of nonzero probability of electron presence around an atom. Values for individual atomic radii can differ depending on the atom's physical and chemical state. Scientists recognize the following atomic radii:

1. Distance between atoms in salts (ionic radius)
2. Distance between atoms in molecules (covalent radius)
3. Distance between atoms in metals (metallic radius)
4. Distance between atoms in condensed state (Van der Waals radius)
5. Calculated radii from theoretical models

Atomic radii do not increase in direct proportion to increasing numbers of protons, neutrons, and electrons. There are many exceptions, but generally for a given configuration of outer electrons, the atomic radius increases with increasing atomic number (moving down vertically on the periodic table), and the atomic radius decreases as electrons are added to any given numbered outer electron orbital (moving horizontally on the periodic table). The latter effect is due to the increased attractive force of the increased nuclear charge. Some examples of calculated atomic radii are 53 pm for hydrogen, 190 pm for sodium, and 298 pm for cesium. Most elements lie between 100 and 200 pm.

Q 32. *How many places can an electron be around an atomic nucleus?*
A. Infinite

Most elementary science textbooks state that the atom is mostly empty space in which the orbital electrons are seen as rotating around the nucleus. However, this assumes that the electron can only be at one place at a time and that it behaves as a point particle. Quantum theory tells us that the electrons are not ever present in any one place, but always have a nonzero probability of being found at any location. Therefore, although counterintuitive, the answer to the above question has to be infinite.

Consider the analogy of a rotating stick to better understand this concept. When stationary, the stick occupies a small fraction of its rotational area, but while rotating at high speed it seems to occupy 100% of the area. We also have to understand that even though the space around the nucleus can be considered "empty" by macroscopic standards, at the quantum level there is a very low probability of another atom's electrons or nucleus occupying that space. This is the reason things appear solid, and we cannot walk through walls. When two substances get close enough to each other, the negative electron clouds of the atoms do not want to enter each other's orbital space, and they repel each other.

However, atoms do at times want to be closer to another atom's outer electrons to form chemical bonds. The science of chemistry is based on this weird property of the electrons wanting to join electrons of other atoms to produce stable molecules under certain sets of conditions, and not wanting to be anywhere near another atom's electrons under different conditions.

Q 33. *What is the size of the average atomic nucleus?*
A. 10 femtometers (10 x 10^{-15} meters)

The atomic nucleus was discovered by Ernest Rutherford during an experiment in which a beam of alpha particles was directed at a thin sheet of gold foil. Most particles passed through the foil, but sometimes one would bounce off at a very sharp angle, indicating that atoms consisted of very dense centers surrounded by large amounts space that high-energy alpha particles could readily penetrate.

We now know that the atomic nucleus consists of positively charged particles called protons of mass 1.67 x 10^{-27} kilograms and neutral particles called neutrons of slightly higher mass. Because of the proximity of positive charges on the protons, a force much stronger than the electrostatic force exists at small distances to prevent the protons from pushing each other apart. This force is called the strong nuclear force. However, because the strong force acts at such small distances, there is a limiting nuclear size above which the nucleus becomes unstable and wants to break apart. That limit is considered to be 208 nuclear particles and is reached at lead-208. Isotopes above atomic weight 208 become increasingly unstable and short-lived.

The smallest atomic nucleus—that of a hydrogen atom—is thought to be 1.75 fm in diameter. The diameter of the nucleus in the much larger uranium atom is 15 fm.

Q 34. *What is the size of a quark?*
A. Less than 1 femtometer (1×10^{-15} meters)

In an experiment conceptually similar to Rutherford's experiment, scientists aimed a beam of high-energy electrons at a proton. They found that unlike the Rutherford experiment, most electrons appeared to go right through the proton, with some scattering at small angles. They concluded that the proton was not a point particle but a smear of electric charge approximately 1 fm in diameter. Scientists Murray Gell-Mann and George Zweig independently theorized that protons and neutrons are made up of three smaller particles, which were named quarks.

Six types of quarks exist; they are described as *up*, *down*, *top*, *bottom*, *strange*, or *charm*. The up quark has a +2/3 charge, and the down quark has a -1/3 charge. A proton is composed of two up quarks and one down quark, and a neutron has one up quark and two down quarks. Quarks also come in three primary colors (metaphorically!): red, yellow, and green. The three quarks in nuclear particles must be all different colors for stability.

Since quarks do not exist in isolation, it is very difficult to determine their size. Based on experimental and theoretical considerations, most physicists consider them to be point objects. The three quarks in nuclear particles are held together by the strong nuclear force. They exchange a force carrier particle between them known as a gluon. They also participate in one of the other fundamental forces of nature known as the weak force, in which quarks can change form and allow the emission of electrons or positrons from the atomic nucleus by means of a short-lived force carrier particle known as a W boson.

Q 35. *What is the size of a string?*
A. 1.6×10^{-35} meters

Isaac Newton's description of gravity works for objects in normal human experience. For entities of cosmological significance like stars, galaxies, and black holes, Einstein's general theory of relativity provides a better understanding of how gravity can actually curve space-time. However, at the quantum level we cannot distinguish between particles and waves, and Newton and Einstein's theories are not adequate. Twenty-first-century physicists are working hard to understand how gravity works at the quantum level and how general relativity can be reconciled with quantum mechanics. This merging of all fundamental forces with general relativity has been called the "theory of everything."

String theory is the most popular current idea for this theory of everything. The basic entities of string theory are not point particles but strings that can form a closed loop or be open with two distinct endpoints. The size of a string is postulated to be 1.6×10^{-35} meters, which is called the Planck length. Vibrations of the strings at different frequencies correspond to the unique elementary particles, similar to the way varying vibrations of strings on a guitar correspond to different notes. Since a string has length, any interaction between two strings occurs over space and time. This effect mathematically eliminates the infinities that result from calculations that would occur from true point particles.

Just like relativity theory, string theory has weird consequences. A mathematical necessity of string theory is that strings exist in more than the recognizable four dimensions of space-time (three for space, one for time). These extra dimensions can curve on themselves and be smaller than the space thought to be occupied by a quark.

Q 36. *What was the amount of excess matter compared to antimatter formed at the big bang?*
A. One part per billion

All of the elementary particles mentioned so far are part of the Standard Model of particle physics. One entire group of particles not emphasized is antiparticles, which have opposite charges and magnetic moments to the known particles. For example, the anticounterpart of the electron is the positron, and the proton has a negatively charged antiproton. There is also an antineutron. Antiparticles can combine like regular matter to create antimatter. Physicists have looked but have not been able to find large collections of antimatter anywhere in the universe. At present, antimatter is known to exist only in experimental situations.

When matter and antimatter particles meet, they annihilate each other and produce energy. An unsolved question of the big bang theory and the Standard Model is why the universe is all matter. During the early big bang, the hot energy should have created equal numbers of matter and antimatter particles, as occurs presently when energy converts to matter. Both types of particles in the early universe should have collided and produced energy. In this circular scenario, the universe should be all energy and no structure. The formal name for this problem is *violation of the charge-parity symmetry*. There is no one theory that comfortably explains this issue, but it is believed that in the very early universe some slight violation of symmetry occurred that allowed for the creation of a slight excess of matter particles on the order of one extra matter particle per billion. This slight excess went on to produce the stars, planets, galaxies, and all the elements of our known universe.

Q 37. *How many elementary forces are there?*
A. Four

In the early universe, the four fundamental forces were united. As it cooled, individual forces became distinct.

Gravity: Weakest force, always attractive, defined by the GM_1M_2/R^2 equation where G is the gravitational constant, R is the distance between two objects, and M_1 and M_2 are the object masses. This relationship breaks down at very large distances (general relativity applies) or at very small distances (the unproven theory of quantum gravity applies). Force is mediated by the hypothetical graviton particle moving at the speed of light.

Electromagnetic force: Second-strongest force, can be attractive or repulsive, and has a similar inverse square relationship to gravity. Force is mediated by photons. All of chemistry is based on this force.

Strong nuclear force: Strongest of all forces, holds atomic nuclei together, and acts at small distances. When holding quarks together, it is strongest and does not weaken with increasing distance. When holding protons and neutrons together, it diminishes with distance, is weaker, and is called residual nuclear force. Its carrier is the gluon.

Weak nuclear force: Force that is the hardest to understand. It works at very small distances within particles and is responsible for some types of radioactivity. It is mediated by carriers called W and Z bosons. These force carriers, unlike others, are massive and are created out of nothing for a minute fraction of a second to allow the reaction to take place. Their existence is very improbable. The weak force stabilizes nuclear reactions within the sun and causes the emission of neutrinos. It is the second-weakest force.

Q 38. *What is the temperature of the hottest flame?*
A. 4990°C

A fire is a chemical reaction between oxygen and carbon and/or hydrogen. It illustrates the energy dynamics involved when atoms and electrons rearrange themselves to acquire the lowest energy state. The unbalanced equation for a complete combustion fire is:

$$(C\text{-}H) + (O\text{-}O) + \text{activation energy} = (O\text{-}C\text{-}O) + (H\text{-}O\text{-}H) + \text{energy}$$

A small amount of heat or activation energy is required to initially break the carbon-hydrogen and oxygen-oxygen bonds. The atoms then rearrange themselves to a more preferred situation for the outer electrons. The result is a more stable chemical bond. The release of energy provides more activation energy, and the process is self-sustaining. Under nonideal conditions where oxygen is limited, carbon monoxide and soot are combustion products.

The visible part of a fire is the flame. It is a mixture of glowing gases and fine particles that emit electromagnetic radiation. The heat released from fire raises the electrons in some of the atoms to higher energy levels. When the electrons fall down to their stable levels, photons are emitted in the visible, infrared, and sometimes ultraviolet wavelengths.

The temperature of a flame depends on the material burning and the oxygen supply. The highest recorded flame temperature was produced by the combustion of the carbon-nitrogen liquid C_4N_2 in oxygen (4990°C)- a rare example where carbon and nitrogen, not carbon and hydrogen burns. Here are some examples of common flame temperatures: candle (1000°C), Bunsen burner (1300°C), oxygen-hydrogen torch (2800°C), oxygen-acetylene torch (3480°C).

Q 39. *What is the temperature at the center of the sun?*
A. Fifteen million degrees Celsius

Hydrogen and helium were created in the early universe. These gases condensed under the force of gravity to form stars. There is a range of star sizes within the universe (from .08 to 265 solar masses). The larger the star, the hotter its center, the bluer its emitted light, and the shorter its life.

Our sun is an average-sized star with an internal temperature of fifteen million degrees Celsius and a life span of about nine billion years. The energy produced by the sun and all stars comes from nuclear fusion. The young sun (present-day sun) will fuse hydrogen nuclei into helium nuclei until all the hydrogen is exhausted. Then the sun will collapse on itself and develop a core temperature of one hundred million degrees Celsius, which can fuse three helium nuclei into carbon. The sun will expand beyond the orbit of Mercury and evolve into a red giant. In time the sun will lose its outer gas layer and collapse around its carbon core, changing into a white dwarf the size of Earth. After billions of years, it will eventually cool to the point where it becomes a black dwarf.

Larger stars develop internal temperatures high enough to fuse atomic nuclei into elements up to iron. In very massive stars, at the end of iron nucleosynthesis the star collapses very quickly and then explodes into a supernova, resulting in very extreme temperatures and energy release. At this point all of the elements of the periodic table—like platinum, silver, and uranium—are produced by nuclear fusion. All of these elements are hurled off into nearby space by the tremendous force of the supernova explosion.

Q 40. *How many chemical elements are there?*
A. 118

An element is defined by its atomic number, which is the number of positively charged protons contained in its nucleus. This number is electrostatically balanced by the exact number of negatively charged electrons in orbitals (probability distributions) around the nucleus. The orbitals are numbered 1 through 7 and have suborbitals within them labeled with letters and containing the following maximum number of electrons: s (2), p (6), d (10), and f (14). As the atomic number increases, electrons fill their orbitals in the following unorthodox sequence of energy levels: 1s < 2s < 2p < 3s < 3p < 4s < 3d < 4p < 5s < 4d < 5p < 6s < 4f < 5d < 6p < 7s < 5f < 6d < 7p.

At the time of the publication of this book, there were 118 known elements. Ninety-eight elements occur naturally on earth, either in a stable or metastable form. With the exception of two elements, atomic numbers 1 through 82 are stable. Atomic numbers 83 through 98 are radioactive and can decay into other elements. Atomic numbers 99 through 118 do not occur naturally in the universe. They are so short-lived that they only exist in nuclear reactors. However, isotopes of elements 120 and 126 may be stable.

Elements are classified as metals, nonmetals, and metalloids. Metals conduct heat and electricity well, have a characteristic appearance, and have chemical bonding in which the outer orbital electrons are free to move around the crystal structure of the whole metal. Nonmetals are poor conductors of electricity and, with the exception of the inert gases, bind covalently to each other. Metalloids (e.g. boron, silicon, arsenic) are semiconductors of electricity and have properties between metals and nonmetals.

Q 41. *How much uranium is required to make a nuclear weapon?*
A. 15 kilograms (kg)

Energy can be released by both nuclear fusion (joining of nuclei) and nuclear fission (breaking apart of nuclei). During fusion, a large amount of energy must be present initially to overcome the electrostatic repulsion of the protons. These conditions naturally exist in the interiors of stars. Fusion can occur up to atomic number 26, which is iron. The nuclei of elements greater than atomic number 26 are increasingly unstable and want to undergo fission. Energy produced by fusion is greater than that produced by fission.

The basis for the first atomic bomb was fission of uranium by a chain reaction. The reaction was initiated by a substance that generated neutrons, which split the uranium nuclei and released energy and more neutrons. When a certain amount of uranium was present (critical mass), the reaction became self-sustaining, and a massive explosion ensued. In this first fission bomb, two pieces of subcritical mass fissionable material were kept separate until the time of detonation, when they were thrust together by conventional explosives. The amount of uranium 235 used was thought to be about 64 kg. Uranium 235 is only present in minute amounts in nature and must be meticulously separated from the more common uranium 238. Now, by improving on the initial design, it is stated that a bomb can be made from as little as 10 kg of plutonium 238 or 15 kg of uranium 235. Such material would occupy a volume slightly larger than a regulation softball.

In a fusion bomb, the energy from a fission bomb is used to fuse hydrogen. Such a bomb would have an order of magnitude of more destructive power than a fission bomb.

Q 42. *What is the lowest temperature?*
A. -273.15°C

In order to understand why there is a lower limit on temperature, one must comprehend what temperature means at the atomic and molecular level. Atoms that are not bound to each other, like helium gas, are colliding with each other and are in a constant back-and-forth motion called translational motion When atoms are bound to each other—like the two atoms of oxygen in the O_2 molecule—the atoms can move back and forth across their chemical bond like a vibrating spring, and also can rotate around some point in the middle of their chemical bond. These two additional types of motion are called vibrational and rotational. All three types of motion contribute to the energy of an individual molecule and the overall temperature and entropy (number of possible microstates in a system) of a group of molecules. In a system of molecules, there is a statistical distribution of the energies of each molecule. The temperature of the system is directly related to the average energy of all of its component molecules. So when oxygen at 90°C is mixed with nitrogen at 40°C, the oxygen molecules lose translational, vibrational, and rotational energy, and the nitrogen molecules gain all three energies. The final temperature of the mixture is somewhere between 40°C and 90°C, and the entropy of the system actually increases.

There is a theoretical temperature at which atoms and molecules have no translational, vibrational, or rotational energy, and have minimum entropy. It is called absolute zero and is -273.15°C. At absolute zero, atoms are technically still in motion because they are subject to the quantum mechanical variations in the position of their electrons, and thus they can change their position. Scientists have come very close to, but have never achieved, true absolute zero.

Q 43. *At what temperatures do rocks melt?*

A. 600–1200°C

All rocks are made of minerals. A mineral is a uniform, crystalline, solid inorganic substance with a distinct chemical composition. Below the earth's crust there is a layer of semifluid rock called the mantle. Although there are high temperatures in this region, there are also high pressures. These high pressures keep the rock in a solid but fluid state. At volcanos there is a connection between the mantle and the surface. During an eruption the mantle rock becomes exposed to the lower pressures at the earth's surface. It becomes a liquid called magma. This liquid contains many different types of atoms in a charged or ionized state. The percent composition roughly corresponds to the percent composition of elements in the earth's mantle and crust. Dissolved oxygen, gas, silicon, iron, and magnesium are common elements in magma.

When magma cools, minerals crystalize out in a characteristic order known as the Bowen reaction series. At 1200°C, the first minerals to crystalize out are olivine and calcium-rich feldspar. At lower temperatures minerals such as pyroxene, amphibole, biotite mica, potassium- and sodium-rich feldspars, and muscovite micas form. Finally at 600°C, remaining silicon and oxygen atoms combine to form quartz crystals. Large non-reactive atoms like silver and gold do not fit well in silica crystal structures and tend to precipitate out. The final appearance of the rock is determined by the rate of cooling. Underground rocks cool slowly, allowing time for atoms to organize themselves into large mineral grains. Granite is an example. Rocks that cool at the earth's surface cool rapidly, have less time for atomic organization, and have smaller grains. Basalt present on the ocean floor is an example. Very rapid cooling can form a glass-like obsidian.

Q 44. *How much does a continent move in a year?*
A. 1–2 centimeters

One of the most amazing discoveries of the twentieth century was the realization that the continents are not fixed to the earth, but are moving on average about 1–2 cm a year. This concept has an interesting history. It was first proposed in 1596 by Abraham Ortelius, the creator of the first world atlas. It was further refined in 1912 by Alfred Wagner who first used the term *continental drift*. While the shape of the continents, their similar geology, and their similar plant and animal life all suggested that the continents were once joined together, there was no convincing theory to explain why the continents would move. Wagner's hypothesis was mostly rejected by the geologic community until the mid-1960s when the theory of plate tectonics was proposed.

It is now thought that the hot and fluid area below the earth's crust acts as a sort of conveyor belt that slowly moves the continents. The earth's crust is divided into a series of plates that slide over this area. This motion results in a pattern in which continents are joined together and then broken up, with old oceans disappearing and new oceans forming. It is postulated that all the continents were joined together approximately two hundred million years ago into a supercontinent known as Pangaea. A theory of repetitive supercontinent formation and breakup in a five-hundred-million-year cycle has been proposed by J. Tuzo Wilson. According to this theory, we are still in the breakup phase of Pangaea, which will last another fifty million years. Then the process will reverse, producing another supercontinent in about three hundred million years. The Cambrian explosion happened after the breakup of the supercontinent Gondwanaland.

Q 45. *How deep is the ocean?*
A. 35,800 feet (≈1 mile above Mount Everest)

The deepest ocean region is the Mariana Trench in the Pacific, which is 35,800 feet deep. About 70% of the earth's surface is covered by ocean water, which contains about 3.5% mineral salts. Minerals come into the ocean by three principal mechanisms: (1) dissolution of rocks by rainwater; (2) volcanic eruptions that spread salts into the atmosphere, which fall into the oceans; and (3) dissolution of minerals in the ocean floor at hot hydrothermal vents. Remarkably, the saline content of the oceans has remained relatively constant over time. The rate of salt loss through evaporation in basins and ocean sprays has balanced the rate at which salt enters the oceans.

Scientists still do not agree on the origin of earth's water. Originally it was proposed that the water "outgassed" from rocks after the cooling of the once-hot earth. It was thought that as the earth's atmosphere cooled, the planet was able to hold water vapor that would eventually fill the oceans with unending rain. More recent theories postulate that terrestrial water came from an extraterrestrial collision with either a comet or asteroid. Radioisotope analysis suggests an asteroidal origin.

It is thought that Mars and Venus may have once had water oceans. There is evidence for hydrocarbon lakes on Titan, a moon of Saturn. It is theorized that other planets and moons in our solar system may have lakes or oceans composed of ammonia, water, or other chemicals. Some speculate that extrasolar planets must exist that are completely covered by water oceans. If such a planet were a few times larger than Earth and closer to its star, and if its oceans were deep enough, there could be a fascinating mix of ice-like material at the bottom and superheated water at the top with bizarre weather.

Q 46. *What percentage of the world's freshwater is stored in glaciers?*
A. 75%

The oceans contain 97.2% of the world's water. Of the 2.8% of the world's freshwater, roughly 75% is present in glaciers, 24% is present as groundwater, and less than 1% is present in lakes, rivers, and streams.

A glacier is composed of ice from compacted snow. Newly fallen snow evaporates slightly and then recrystallizes into sand-sized pieces of granular ice. The pressure from years of accumulated snow melts the granular ice at the surfaces of the crystals, producing water that fills in the spaces. The result is a solid mass of glacial ice. Snowcaps are not considered glaciers since they melt during the summer.

There are two types of glaciers—continental and alpine. Continental glaciers are on top of the land masses of Antarctica and Greenland. They were also present in other parts of the world during previous ice ages. Alpine glaciers cover the tops of mountains.

Both types of glaciers are in motion. Alpine glaciers can move downhill at the rate of a foot a day. Continental glaciers move by the pressure from the snow that is almost constantly falling. They move at a slower rate of fifteen feet per year.

Both types of glaciers result in land erosion by breaking the underlying rock mass and lifting the pieces into the ice to act as an abrasive. Alpine glaciers erode U-shaped valleys. In contrast, continental glacial erosion can actually raise the elevation of the landscape. Many current lakes were created when large ice chunks broke off from continental glaciers present during previous ice ages.

Q 47. *How old is Antarctic ice?*
A. 780,000 years

Ice in the Antarctic glacier is over two miles thick. Each year a new layer of snow is added, which compresses the previous layers. At the top, annual deposits can be distinguished like tree rings. Near the bottom, the ice is under so much pressure that samples reflect average time periods. Scientists can sample ice cores down to the bottom of the glacier, and thus have an unbroken sample of ice from various times in Earth's climatic history. Radioisotopic dating tells us that ice at the bottom is 780,000 years old.

The chemical composition of the layers reveals the climate for that period. Analysis of trapped air bubbles tells us about the chemical makeup of the earth's atmosphere for a period. The industrial revolution is reflected by a large increase in the amount of carbon dioxide present from 1800 onward. Since oxygen-18 is heavier than oxygen-16, and since it would require higher temperatures to evaporate oxygen-18 compared to oxygen-16, the ratio of these two isotopes present in the ice is directly related to Earth's temperature. Also, over hundreds of thousands of years, there is a direct correlation between carbon dioxide concentrations in air bubbles and temperature.

Evidence from the ice cores demonstrates that the earth experiences a cycle of extreme coldness and warming at intervals of about every 100,000 years. This change causes a melting and reformation of glaciers and changes in continental shorelines. Cyclic changes in the earth's distance to the sun (100,000-year cycle) and rotation axis (41,000-year cycle) contributes to this 100,000-year temperature cycle.

Q 48. *How many years did it take to produce the world's oil and coal supply?*
A. 350 million years

The world's petroleum was produced from small marine algae and zooplankton that settled to the bottom of ancient oceans in sandstone or limestone reservoirs. With time and pressure, the material decomposed into a waxy substance called kerogen, which eventually transformed into petroleum and natural gas. A rock formation (ideally shale) caped the decaying organic material and preserved the oil and gas until the present time. The majority of the oil produced was not caped, and it made its way to the surface where it was consumed by oil-eating microbes. Most of the world's oil comes from the Middle East. The United States consumes about 30% of the world's oil, but only has 2% on its land. The global consumption of oil is thirty-one billion barrels a day. It is estimated that we may have as little as thirty years left in reserves.

Coal comes from wood that has settled in an area of low oxygen, low microbial activity, and high acidity. Wood that has been preserved at the bottom of a swamp is called peat. If peat is subjected to pressure, it becomes coal. Coal is rated by its carbon content. The percentage of carbon in various forms is as follows: wood 45%, peat 55%, lignite 65%, bituminous 75%, and anthracite 95%.

It is impossible to precisely tell how many years it took to produce the world's oil supply, but it probably was longer than that for coal. A reasonable estimate for coal is that it began with the first forests about 350 million years ago. A staggering fact is that what nature took 350 million years to remove (carbon from atmosphere in the form of carbon dioxide) mankind is putting back in about 300 years.

Q 49. *What is the energy density of gasoline?*
A. 34 megajoules/liter

Since the world may run out of oil before the middle of the twenty-first century, scientists are frantically trying to find a substitute for gasoline. The ideal replacement should have around the same cost per energy density. Energy density is the amount of energy released when a fuel is burned under ideal conditions per volume. The following table depicts the energy densities for some common fuels:

Fuel	Energy Density (MJ/L)
Diesel	38
Gasoline	34
Ethanol	24
Hydrogen under pressure	7
Natural gas under pressure	10
Lithium ion battery	2

The advantages (A) and disadvantages (D) of the alternatives to petroleum-based diesel and gasoline are discussed below.

Ethanol: A—Unlimited supply from plants; D—requires water and diverts land that could be used for food production; not cost-effective.

Pressurized hydrogen: A—no carbon emissions, unlimited source from seawater; D—needs electricity to produce, not cost-effective.

Pressurized natural gas: A—cheap, better for engines than gasoline; D—must be stored under pressure and eventually will be depleted.

Lithium ion battery: A—cheap to recharge, not dependent on fossil fuel; D—low energy density, safety issues, high initial cost, longevity.

Q 50. *How many transistors can be placed on a computer chip?*
A. Three billion

The first general-purpose electronic computer was developed in 1946. It used vacuum tubes but processed information digitally. Other computers at the time used measured data and were called analog computers. Digital information is coded by a series of zeros and ones. Any piece of information can be processed digitally. For example, each musical note can be represented by a unique series of zeros and ones. The development of the transistor by Bell Labs in the late 1940s shifted computers toward the digital mode. A transistor is a small electronic device that either opens or closes an electronic circuit. It replaced the larger and more fragile vacuum tubes.

In 1965 Gordon Moore, the cofounder of Intel, predicted that the number of transistors that could be placed on a computer chip would double approximately every twenty-four months. This law has been surprisingly accurate up until recent times. Eventually, it will be limited by the size of individual atoms. At present, an amazing three billion transistors can be placed on a single computer chip.

Scientists are working on alternative ways to extend Moore's law:

Three-dimensional chips: These can have a higher transistor density and have greater efficiencies, but they generate more heat.

Optical computing: These computers use photons instead of electrons to process information.

Quantum computing: This is investigational and nonbinary, based on the quantum mechanical properties of atoms; it could be faster than transistor chips.

References

Some of the information in this book was obtained by listening to the presentations in the following DVD courses produced by The Great Courses, Chantilly, Virginia. (www.thegreatcourses.com). I highly recommend these lectures to anyone who wants a deeper understanding of the subject matter.

Big History: The Big Bang, Life on Earth, and the Rise of Humanity. David Christian, D. Phil.

Understanding the Universe: An introduction to Astronomy, 2nd Edition. Alex Filippenko, Ph.D

Cosmology: The History and Nature of Our Universe. Mark Whittle, Ph.D

Dark Matter, Dark Energy: The Dark Side of the Universe. Sean Carroll, Ph.D

New Frontiers: Modern Perspectives on Our Solar System. Frank Summers, Ph.D

Joy of Science. Robert M. Hazen, Ph.D

Particle Physics for Non-Physicists: A Tour of the Microcosmos. Stephen Pollock, Ph.D

Einstein's Relativity and the Quantum Revolution: Modern Physics for Non-Scientists, 2nd Edition. Richard Wolfson, Ph.d

Nature of Earth: An Introduction to Geology. John J. Renton, Ph.D

How the Earth Works. Michael E. Wysession, Ph.D

Major Transitions in Evolution. Anthony Martin Ph.D and John Hawks, Ph.D

Origins of Life. Robert M. Hazen, Ph.D

Understanding the Science For Tomorrow: Myth and Reality. Jeffrey C. Grossman, Ph.D

Superstring Theory: The DNA of Reality. S. James Gates Jr., Ph.D

The Physics of History. David J. Helfand, Ph.D

Impossible: Physcis beyond the Edge. Benjamin Schumacher, Ph.D

 Other References

Gronenberg W. Structure and function of ant (Hymenoptera: Formicidae) brains: Strength in numbers. *Myrmecol. News* 11: 25-36 (online 3 May 2008)

Aiello P, Wheeler P. (1995) The expensive tissue hypothesis: The brain and the digestive system in primate evolution. *Current Anthropology* 36 (2): 199-221.

Wessner DR (2008). The origin of viruses. *Nature Education* 3 (9): 37-39.

Beef industry's approach to controlling E. coli. www.explorebeef.org

United Nations World Population Prospects: 2006 Revision

Mora C, Tittensor DP, Adl S, Simpson AGB, Worm B, (August 2011). How many species are there on earth and in the ocean? www.plosbiology.org

Dean T. (22 August 2006). Uranium enrichment: How to make an atomic bomb? Cosmos online. www.cosmosmagazine.com

An overview of the human genome project.
www.genome.gov/12011238

Cotton FA, Wilkinson G. (1988). *Advanced Inorganic Chemistry* (5th ed.) Wiley, 1385.

Klaunig JE, Kamendulis LM. "Chemical Carcinogens", in *Casarett & Doull's Toxicology: TheBasic Science of Poisons*, 7th ed., ed. CD Klaassen (New York: McGraw-Hill, 2008), 329-380.

Wilcox AJ et al (1988). Incidence of early loss of pregnancy. *N Eng J Med*, 319 (4): 189-194.

Kirshenbaum AD, Grosse AV (1956). The combustion of carbon subnitride C4N2 and a method for the production of continuous temperatures in the range of 5000-6000 °K. *Journal of the American Chemical Society* 78 (9): 2020.

About the Author

Dr. Richard J. Fruncillo spent most of his professional career conducting clinical research studies for major pharmaceutical corporations. In addition to a medical degree, he has an undergraduate degree in chemistry and a PhD in biochemical pharmacology. He is certified by the American Board of Internal Medicine, the American Board of Clinical Pharmacology, and the American Board of Toxicology. He has previously taught pharmacology and toxicology at the medical and graduate school levels, and he is the author of many scientific publications in the areas of basic and clinical pharmacology and toxicology. Currently he is a consultant to the medical and legal professions in the areas of pharmacology, toxicology, and drug development.